Dr. James Harris

Dr. James Harris is a professor of supply chain and quality Management in the UAE. He has lived in the UAE for six years but has been in the Middle East for over 17 years. James is an American who spent four years in France earning his Doctorate of Business from Grenoble Ecole de Management in France and both his Master of Business and his Finance Bachelors in the US.

As a creator, James has undertaken several entrepreneurial ventures, including being the founder and creator of the Heritage Lemonade brand as well as being a major beverage supplier for the US Department of Defense. In the Middle East, he established 'The Business Doctor,' a training and consultation company in Kuwait. Dr. James has developed an athletic cryo-suit to prevent post-workout swelling (Patent pending). You can learn more about Dr. James Harris on the below podcasting and video links:

Subject	Link
Epic Failures in Entrepreneurships	https://www.youtube.com/watch?v=jH_UfnR1Wpc&t=1070s
Dubai Chats w/ Quality Management Expert Ihab Hussein	https://www.youtube.com/watch?v=u3Tq2Bm_SPI&t=183s
Dubai Business Chats with Supply Chain in the UAE: The Steak Case	https://www.youtube.com/watch?v=1lm2_nfR86M&t=3s
Prototype Process: Project Freeze	https://www.youtube.com/watch?v=Qch3fQnx6_Q&t=7s

I dedicate this first book to my three wonderful and grown-up children, Tawney, Chuck, and Maximus, for the many years that I was absent from the family scene in pursuit of my proverbial 'hunt for life's treasures.' You have all turned out to be great entrepreneurs and independent contributors in your own communities. You have now become the role models of tomorrow. For this, I am your fanboy.

Secondly, I dedicate this book to my student family in the UAE. You, the Quality Management and Entrepreneurship group, make my work environment a true pleasure.

Not only you, but our staff and faculty are the happy garden that allow us to grow mentally and spiritually in a time where the global workplace has become a desert of stress and depression. We still have fun and for that, I thank you all.

James Harris

IDEAS TO INVENTIONS

The Art of Turning
Ideas into Reality

AUSTIN MACAULEY PUBLISHERS™
LONDON • CAMBRIDGE • NEW YORK • SHARJAH

Copyright © James Harris 2024

The right of James Harris to be identified as author of this work has been asserted by the author in accordance with Federal Law No. (7) of UAE, Year 2002, Concerning Copyrights and Neighboring Rights.

All rights reserved. No part of this publication may be reproduced, stored in a retrieval system, or transmitted in any form or by any means, electronic, mechanical, photocopying, recording, or otherwise, without the prior permission of the publishers.

Any person who commits any unauthorized act in relation to this publication may be liable to legal prosecution and civil claims for damages.

The age group that matches the content of the books has been classified according to the age classification system issued by the Ministry of Culture and Youth.

ISBN – 9789948778653 – (Paperback)
ISBN – 9789948778646 – (E-Book)

Application Number: MC-10-01-9092468
Age Classification: E

Printer Name: iPrint Global Ltd
Printer Address: Witchford, England

First Published 2024
AUSTIN MACAULEY PUBLISHERS FZE
Sharjah Publishing City
P.O Box [519201]
Sharjah, UAE
www.austinmacauley.ae
+971 655 95 202

I would like to express my deep and sincere gratitude to Ms. Preeta Pais for being a super 'Word Smith.' I extend my heartfelt thanks for listening to my many crazy ideas over the years and then, helping develop many of those ideas into tangible projects. You are the greatest, Preeta.

I would like to say thanks to my family and friends who support me by letting me use them as cases within the book. Dr. Chara Lee-Brinson and Sharon Nakitto, your case was excellent and heartfelt. Thank you both for letting me interview you. In addition, great appreciation to Dr. Shahira Osama. You are the epitome of a solid leader and could take over the world if you want.

Lastly, I must acknowledge the leadership at Ras Al Khamah Colleges in the United Arab Emirates. You continually create a family environment where we can aspire to be better than we are. Below are a few of my university family that create a magnificent environment in which we spend the vast majority of our days:

Maryam Alhaffeet
Mouza Alshemaili
Randa Alzaabi
Kwashawn Barnett

Khwaja Khan
Emad Masoud
Samantha McDonald Amara
Muna Ahmed
Kennedy Modugu
Juan Dempere
Eseroghene Udjo
Marilou Maderazo
Pranav Kumar
Sabir Malik
Aaesha Alshehhi
Alya Alhebsi
Bharathan Viswanathan
Ebrahim Aleamash
Fatema Alali
Fatema Alhebsi
Giovanna Bejjani
Iva Bulatovic
Marisse Aranas
Mohamed Salem
Mouza Alkhanbouli
Moza Almazrouei
Muhammad Qamar
Sumayya Rashid
Sean Quinn
Gerry Gibson
Ihab Hussein
And, Suaad Al Mansoori, the person that inspired me to write for myself

Table of Contents

Abstract	**11**
Preface	**13**
Introduction	**17**
Chapter 1	**23**
1.Levels of Prototype Fabrication	*23*
A. Low-Fidelity Prototypes	23
B. High-Fidelity Prototypes	27
C. Feasibility Prototypes	29
D. Live Data Prototypes	29
E. Other Types of Prototypes in Software	30
Chapter 2	**31**
2. Technical Design Phase	*31*
Chapter 3	**35**
3. Fabrication Phase	*35*
A. Fabrication Choices	*35*
Case – From Cancer Research to Hair Treatment	*41*
Chapter 4	**44**

4. Packaging and Logo Design		*44*
A.	Packaging Overview	45
I	Generic Steps for Packaging	45
II	Logo and Packaging Draft	48
III.	Developing a Mock-Up	50
IV	Package Marketing Test and Feedback	52
Case – Interview with a Graphics Expert		*52*
Chapter 5		**57**
5. Prototypes Testing		*57*
F.	Simple Steps for Testing a Prototype	59
Chapter 6		**65**
6. Intellectual Property Phase Overview		*65*
A.	How to File a Patent in the United Arab Emirates	66
B.	General Mandatory Documents Required	68
Chapter 7		**70**
Bibliography		**82**

Abstract

Creating an idea prototype is pivotal in devising an accurate feasibility picture with realistic financial implications. A prototype is a visual, tangible representation of the business idea and is essential to creating trusting and get honest feedback on the product. In addition, physical models help formulate a better business plan by outlining actual product costs. Prototyping gives the inventor a sense of clarity. (www.forbes.com)

After the initial originality of a product business concept, the next step is the creation and commercialization of prototypes, which is significantly understudied. (Kunicina et al 2019).

The chapter aims to provide a standard yet flexible framework within the prototyping phase that shows sequential milestones for product developers to follow. Although actual steps may vary due to the dynamics of differences in industry needs, this prototype development section focuses on six primary areas, including (1) technical design, (2) fabrication phase, (3) budgeting and trial phase, (4) feedback and modification phase, (5) intellectual property phase, and implementation planning. In terms of scope, this section starts after the product description in the initial business plan and includes a recommendation for intellectual property protection. Ultimately, the areas provide transitions from creative ideas to testable products.

Interested readers may include students, small business novices, non-academic entrepreneurs, and organizational employees with research and product development responsibilities. The chapter offers tools and sequences for product design and development for these groups. These target groups will experience numerous product development cases to improve the reader's skills and competencies in new product development milestones and best practices.

By the time this chapter is completed, readers will be competent in the following areas of product planning: generating product drawings, creating initial products specifications, sourcing help with fabrication, developing a budget for prototypes, testing and obtaining feedback, refabricating and final drafting, and applying for copyright/intellectual property. Ideas alone tend to be inadequate to influence investors, engage targeted audiences, or take on financial risk. Completing a functional prototype is the best illustration, translated into tangible, usable, and sellable products.

Preface

Although the writer is an academic and an entrepreneur, the book is explicitly written for ordinary people with great ideas and the potential to grow the concept who don't know what to do next. This book is about developing visual representations of a belief to succeed. In addition to creating business plans, a prototype needs to be made. 80% of our perception, learning, cognition, and daily activities are impacted or mediated by visual representation. Below is a mildly paraphrased poem from American author Langston Hughes:

I'd rather see the sample than hear of it any day.

I'd rather one walk with me than just to tell the way.

The eye is a better pupil and more willing than the ear.

Advice may be misleading, but EXAMPLE is always clear.

And the very best of teachers are the ones who live their creed,

For to see good put into action is what everybody needs.

I can soon learn to do it if you let me see it done.

I can watch your hand in motion, but your tongue too fast may run.

And the lectures you deliver may be very fine and true.

But I'd rather get my lesson by observing what you do.

For I may misunderstand you and the fine advice you give.

But there's no misunderstanding how you act and how you live.

Live Your Creed by Langston Hughes is one of my favorite poems of all time. It inspired this semi-academic book; however, it is written for regular people. As a chronic entrepreneur, I discovered 35 years ago with my first primary business, the 'product sample' was the prominent variable important people having confidence in my business idea.

The primary purpose of my writing is to inspire those with brilliant ideas looming in their minds to release them from the forest of your mind to humankind's hand. Tradition and popular philosophy dictate that ideas evolve from the design thinking process. This book does not contradict this prevalent thinking but highlights and expands on developing your inventive concept into something people may touch. We are going to refer to this as a prototype throughout the book. The target people of your may include investors, personal critics, family and friends, banks, teachers, and anyone else you would need to influence about how magnificent your product or service idea will be someday.

In the 1990s, I started a juice manufacturing company called 'Heritage Southern Lemonades.' The idea was not an invention, and the concept was quite simple. I had just graduated from the University in Southern Alabama. I also knew that local beverages had a historical and unique taste to people in the southern area of the United States. I had written an extended and extensive business plan and solicited many investors to start the business. However, after 18 months of investor meetings, chamber of commerce visits, and bank presentations, not one target group was interested as a potential client. After multiple epic failures, my daughter asked me, 'What does our lemonade taste like, Dad?' In this eureka moment, I realized that I had focused on the theory of the business planning process and not actually making the best lemonade flavors in the South for everyone to taste.

Within the next six months, my best friend Robert and my kids created over 20 flavors of lemonade flavors and other juices unique to southern American taste buds. Not being an artist, we drew a terrible logo design on the side of a brown bag. With the help of a graphics student at the local university, a prototype and package for Heritage Lemonade were born. We bottled over 5000 samples of our inconsistent 'kitchen-made' flavors. We got enough public feedback to choose our first four flavors: Southern Lemonade, Peach Lemonade, Mint Julep Tea, and Watermelon Lemonade. In 1997, people chose our package design, and our four flavors and investors came from everywhere.

Additionally, they were the first to buy our home-made product. In the end, the problem was that I had initially focused on the theory of my business idea, but I had not created a tangible vision for all to touch and taste. The principle of my story is, 'If you build it, they will come.'

Hopefully, this book offers you, the reader, a sequential guide to taking your creative idea and converting it into a more tangible representation of what is in your head. My attempt in writing this piece is to offer simple prototyping concepts that will eventually develop into a 'proof of concept' and its basis for intellectual property rights. Moreover, the book is easy to read while still immersed in sound entrepreneur-based best practices and academic theory. Because the Middle East has risen to the forefront of Innovation, creativity, and technology, this writing has been paraphrased and translated into Arabic in the book's second part. The cases and examples are similar but more relevant to the Gulf area.

Enjoy your reading; I would love your feedback and hope you receive something new and different from this book.

Introduction

This chapter highlights aspects, best practices, and processes from 'idea to product.' The objective is to develop, test, and adjust the idea's feasibility. Proof of concept (POC), also known as proof of principle in various industries, is the development of a working prototype that can be presented to investors as a tangible visual representation of the start-up's concept. One of the main problems start-ups have with investors is that they do not value start-up ideas/promises without proof that the products or services are feasible and profitable.

Prototypes, proof of principle (POP), and proof of concept (POC) fall within the area of products or services that assist with investor participation. POPs and POCs are prototypes that are ready for potential client testing. POP/POC are not prepared for sale and will not have many of the details of the end product.

The objective of a prototype is testing. The innovator may now solicit users who can discover any flaws, understand the missing functionality, and rate the overall reaction to the user experience. Additionally, the innovator may produce a more effective and functional product or service model that provides an understanding of its architecture, functionality, and/or layout.

Theory to tangible

A prototype is an initial attempt to make a working model usable in the real world. Stuff goes wrong in the process, but the principal aim of constructing a prototype is to recognize such errors and stumbling blocks. A prototype has almost all.

The final POP will not be bug-free but will eventually demonstrate the concept's functionality. Also, proof of concept means that the technological viability will be built and tested when a prototype reveals a possibly flawed, unrefined attempt at the final result. Furthermore, it also needs an all-powerful consumer to assess the product's performance (2020 Kunicina et al).

A working prototype [6] is an early illustration, a product sample intended to evaluate an idea or method from which to practice or demonstrate to customers and marketing firms. A prototype should translate the vision of an invention or product design into something real and tangible. For less measurable goods, such as applications, the look and sound of the final user can be replicated on specific hardware units. Below are listed some key reasons to make a working prototype:

The types of prototypes are as follows:

1. **Service Prototypes – as the name suggests, these prototypes create a real interactive experience to test a new service or modify an existing service.** Service prototypes are drafts of the actual service process experience intended to examine and validate design choices. They are working models that offer

most of the product's functionalities and are used as a stepping point to the final product. Service prototypes are not ready to be released for business use and are used only for testing purposes.

When the initial draft is delivered, the prototype becomes a **service proof of principle**

(POP)/service proof of concept (POC).

2. **Product Prototypes – are tangible drafts of the design idea.** They are partially and/or fully working models which offer some or most functionalities. Similar to service prototypes, product prototypes are not ready for market release. They are built for initial experimentation and potential reaction within a sample size. Both entrepreneurs and research teams develop prototypes so users can discover any flaws with the product. Usually, market tests are created for users to try the product and provide feedback. Product prototype testing is standard across various industries (e.g., software development, engineering projects, architecture, medical and healthcare product development, etc.).

A product prototype or proof of concept makes more of an impression on investors than write-ups and verbal descriptions.

On the other hand, product prototypes may be exclusively developed for potential client customers and/or investors to gauge their potential in them firsthand. They usually lack the polish of the end product and have limited possibilities.

From an organizational standpoint, there are significant benefits to the basic POP/POC process:

The POP/POC process helps sample the need levels for a potential consumer market. The data from this research allows entrepreneurs and research teams to value the potential investment. Large, established companies spend millions on research testing and technologies to validate their prototypes and develop the projects' potential for success. On the other hand, entirely new companies may test the idea straightforwardly for a fraction of the cost.

1. The POP/POC process demonstrates to senior management and entrepreneurs the feasibility level of the project's function. The theory behind the idea will

often differ when the project operations are exposed. Even the best ideas frequently need improvement. Even if it works, most of the innovators, as well as other stakeholders of the potential or existing companies, may not understand everything about the prototype until they see it, touch it, and examine it.

2. The POP/POC process helps clarify the idea's financial implications. Teams gain a better comprehension of fabrication cost, supply chain cost, procurement cost, set price, design cost, delivery cost, etc. This way, future investors and/or other existing stakeholders may see a more accurate financial picture. Overall, this process generally reduces the risk of failure for potential stakeholders. The investment risks will be dramatically reduced if a product has a validated working prototype (2020, Kunicina et al).

3. The POP/ POC process improves the original prototype design because modifying the design is usually required for various reasons. During the transition from Prototype to POP/ POC, adjustments can come from customer requirements, addition/ deletion of usage functions, aesthetics functions, package design, raw material changes, etc.

4. Last, inventors might need to add or modify their business plan through a business model or consumer target. The POP/POC process would enable innovators to redefine the target market or product location or even terminate the entire project before wasting investor funding at the production phase

later. The proof-of-concept results solely represents the innovator's due diligence.

As an innovator engages each process, their product or service will reduce its chances of function failure or customer acceptance. Within hypothesis development, validation of proposed ideas is extremely valuable. Validation helps the innovator see the idea's functionality and comprehend how the prototype functions in a controlled environment. The product/service has been tested for its required variables, such as consistency and feasibility. The innovators and investors must consider that the samples may not be an exact representation of the ever-changing market at the time of project completion. Time and other external market factors will increase risk. At this point, the innovator may also consider abandoning the project if the experiment hypothesis's results are negative. Ultimately, these essential processes aim to define the product's features before beginning production (2020 Kunicina et. al).

Prototypes show how an idea moves to a functional, sellable product or service. It is an evolutionary process that converts an idea into a storyboard, a storyboard into a prototype, and a prototype into a proof of concept.

Chapter 1

1. Levels of Prototype Fabrication

There is no 'one size fits all' for developing a prototype. A home-baking entrepreneur is more likely to create samples of their food with more focus on the packaging design. A software app developer may generate a dummy app prototype with at least partially functional software to influence investors to fund the project. Conversely, a 'housekeeper android' prototype requires several fabrications (i.e., robotics, electrical & mechanical fabrication, software fabrication). The inventor must choose their approach to prototypes based on the creator's and stakeholders' needs.

Several options are available for developing diagrams, cartoons, 3D fabrications, semi-functional prototypes, detailed rendering, actual samples, and fully functioning fabrication. The levels of prototypes mainly fall under the category – low fidelity and high fidelity.

A. Low-Fidelity Prototypes

Low-Fidelity Prototypes are a quick and easy way to render high concepts into tangible representations. They are mainly in the form of drawings or made of simple materials that describe the general idea rather than the fine details. They do not allow any user interaction. A low-fidelity prototype is a hand-drawn sketch or a mock-up on paper. Alternatively, digital prototypes are created in standard software for prototyping as well. Both types of rendering describe the overall idea, transitions, and basic functionality. At this point, only minimal testing of this type of prototype can be executed. In the Low-Fidelity option, the creator can solicit feedback about user perception of the product, design, or flow.

Photo by www.picjumbo.com

Type of creator	Target	Benefits
• Application and software • Animation Cinema and Film • Marketing & advertising concepts • Business process • Futurist concept • Art and Entertainment	• Potential customers, investors, bankers • Several stakeholders (i.e., employees, Suppliers and Vendors, Communities, and governments)	• Time and cost-effective • Easier to edit the prototype • Helps make the concept of your project clear to users • Low-fidelity prototypes put less pressure on users. They can feel more relaxed and express their views in more detail; • The intermediate stage of the design will be available for stakeholders, so they will not expect the complete design immediately;

a. Drawings, Cartoons, and Storyboarding

This technique involves sketching/drawing your product's various processes, functions, and design details. Since this involves free-hand sketches and paper, it is easy to change and cheaper to create. There is less need for outside consultation. These paper renderings make it easy to describe ideas and explain complex plot concepts, and the process or story layout sequences. This technique is most famously used in the movie business, where the entire story of the movie is sketched on paper and laid out on boards; hence, the term

'Storyboards.' The example below shows the storyboard of the film 'Titanic.'

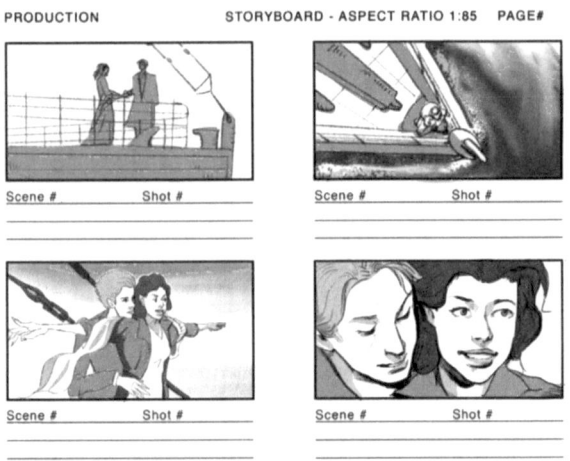

[Storyboard] Titanic (1997)/study (sillyfilmofficial.com)

b. Wireframe Prototypes

Much like storyboarding, a **wireframe** is a digital diagram or layout of the product which can be shifted to suit the design. You can use an app like ***Balsamiq*** to quickly create an illustration of your product. This option is appropriate for civil projects, software/websites, or business processes. A wireframe is a low-cost option that can influence or communicate with developers to navigate the structure and placement of different web content/milestones/process relationships.

c. Video Prototypes

Video prototypes often present a product as an animated video or a simulation that explains and graphically represents a project. The films show other prototypes to help stakeholders – like fellow designers, the management, or even consumers – visualize the product.

d. Samples/Working Prototypes

Samples and working prototypes let the creators test the idea of a product to see if it actually functions the way it is intended. In many industries, creators make samples (i.e., food and beverage) or working prototypes (i.e., electric auto) to test user perceptions or preferences. Models and working prototypes are helpful for textile, mechanized, culinary inventions, or other designs with features that need to function, look, or taste a certain way. The cost tends to be more than storyboarding or drawings; however, user feedback is much clearer. Additionally, the ability to positively influence stakeholders and investors increases significantly.

B. High-Fidelity Prototypes

On the other end of the fabrication spectrum, high-fidelity prototypes tend to be highly interactive, functional, and close to the final product. These prototypes contain all the details and functionality, so they are the most appropriate for testing hypotheses. They are usually computer-based and allow for user interactions.

High-fidelity prototypes are usually created in the later phases of the design when the inventor knows what they have to build and needs to test them on users to better understand consumer/user feasibility. In research and development, these prototypes are built upon existing versions, such as 2.0 and 3.0 improvements related to existing products.

If these prototypes test positively, the risk of user rejection is low, and the project situation is ideal for moving forward. These prototypes are hugely expensive, require expert assistance, and take the most time to fabricate.

Type of creator	Target	Benefits
• Application and software • Restaurants • Transportation	• Potential customers, investors, and bankers • Several stakeholders (i.e., employees, suppliers and vendors, communities, and governments)	• The prototype is exceptionally so detailed, so functionality can be tested. • It is possible to test the hardware components, dropdown menus, filters, and the various input fields • Feedback from users about visual graphics (e.g., animation icons, charts, photos, illustrations) • Able to test commercial aspects of the site (pay gate, web security, client information)

C. Feasibility Prototypes

Feasibility prototypes are created to test a particular function of the product. It may be a specific technological element or precise fabrication, and/or it needs to be tested within the context of a significant project. These prototypes do not require a considerable time or resource commitment. Still, they target significant unknowns, acting to reduce risk and expose a clearer development roadmap based on a proven central proposition.

For example, suppose a heart rate sensor embedded within a fabric was key to a product proposition. In that case, a feasibility prototype may target the integration of a specific electrode into a fabric substrate. This could allow tests to be performed on the functional performance, the mechanical properties, or the washability of the proposed solution.

D. Live Data Prototypes

Live Data prototypes are more specific prototypes that explicitly relate to information systems and software building. Live Data prototypes use data/application programming Interface to insert actual content into a prototype. The purpose of this kind of prototype is to test some programming hypotheses with the use of real data.

E. Other Types of Prototypes in Software

e. Horizontal Prototypes

A horizontal prototype shows a design from the user end and is used in the early stages of analysis. It is mainly used in software design to help engineers understand the human interface of a project by including sample screens, buttons, pop-ups, menus, etc. on a computer to test how users interact with the product. They are suitable for presenting ideas to stakeholders, aiding requirements discussions, and design decisions.

f. Vertical Prototypes

Vertical prototypes are used in the later stages of analysis and are more technical. They are created to better understand complex or specific features. They show a more complex function of the critical elements to prove to the stakeholders that the application works. They are considered 'back end' models used to test essential functions in software before it moves on to another design phase.

Chapter 2

2. Technical Design Phase

Once you have decided on the level of prototype you want to create, you have to begin the technical design. You can start by designing the product on paper or by creating an actual development model. During this phase, an inventor can use various techniques to ensure that the prototype will conform to the expectations of both the producer and the consumer.

1. Paper Rendering

Start by sketching out your product on a piece of paper. Using paper and ink is the best way to spot concept mistakes and redesign the product. Make a rough drawing showing what the product should look like, or write down the features and a description of your prototype. Conversely, you can make a paper prototype with more accuracy and added realism with user interfacing. At this stage of the production, since paper prototypes are more cost-effective, it promotes innovation and replication.

The sketch can incorporate necessary features and can be rudimentary without any extra features just to enable you to decide what it should look like.

You can omit fancy embellishments like voice commands and colors to your robot dog design until you have decided how the technology will function.

This will help you before you invest in an actual prototype.

2. Teardown

A critical factor in the life cycle of any product is competition. Unless entirely original, your product will have competitors in the market. A part of product development is to investigate the competitor's product, which will give you an idea of the 'Why' of the product. The pertinent questions are why the product works and what the assembled parts are.

The best way to investigate the competing product is to get it and tear it apart. If your product is a jetpack, take apart a competitor's jetpack. Note the materials used and the design so you can improve on them in your product. If it is software, read the code, find the bugs, and eliminate them from your software.

You cannot copy the design since that will get you in legal hot water. Instead, use the teardown to learn how to improve your own product.

3. Handmade Version of Your Design

If the product can be made at home and/or in miniature, it will save money, which is an essential factor for a start-up. For example, creating a sample of a leather handbag using discarded nylon clothing will help the consumer see the finished product. Architects usually create mock-ups of the buildings to show investors the final look of the property.

Alternately, using different materials, processes or techniques might help with the design process and open up new avenues or give new ideas.

4. Modeling Program

If the product allows for it, another option is to use a computer modeling program to create a digital 3D interactive prototype to let the customers see how the final product will operate. Many software programs, such as Prototyping on Paper and in vision, allow you to create an interactive model without hiring outside help.

5. CAD (Computer Aided Design)

Once you have an idea of what your product will look like, you can create a CAD of your Prototype, if you cannot make it at home on your own. This will require hiring CAD engineers who will use software that will render 2D or 3D models of your creation. Using a computer will help visualize the product in more explicit detail and helps streamline the designer's workflow.

CAD programs are also used to create virtually simulated prototypes in case you display something like a real estate project or automobiles that otherwise cannot be shown in a smaller space. Virtual prototypes will deliver your product to scale and its actual functions via virtual simulation.

Using CAD also creates a record of the design process, which helps protect the design and is used in the patent application process. The CAD design will allow the prototype manufacturer to get a clear and detailed idea of the final product.

Chapter 3

3. Fabrication Phase

Fabricating a prototype – Fabrication of the innovator's product is a necessary step between computer rendering and production. Before an inventor can begin production, he must create a semi-functional model to know the cost of the invention. In the past, the fabrication of a product has traditionally been limited to simple mock-ups that one could create with simple household materials. Nowadays, inventors can source Prototypes from an array of organizations such as universities, machine shops, 3D printing firms, FabLabs, or web designs and coding. Prototyping has progressed quickly, with many more options with varying costs and complexities.

A. Fabrication Choices

- Step One: Decide on the areas of fabrication required.

The inventor must identify the purpose of the prototype and determine the exemplary fabrication service and technologies to achieve completion. Below are some of the areas that have to be considered when fabricating the prototype:

- Architecture and Construction
- Civil and environmental science
- Food science
- Textile science and fashion
- Computers and Security
- Mechanical, hydro, and electric
- Software and applications
- Manufacturing and business process engineering
- Cooling and heating

After understanding what areas the Innovation includes, there is a litany of choices of fabrication services. Whether you intend to test new technology in your product, recreate the final aspect and aesthetics, or just some specific functionality, each goal's approach is a bit different. Obviously, the options available differ. Some products require consultations with two or areas of fabrication areas. The innovator must consider various product characteristics, which materials will be used, and which fabrication expertise and technologies should be used to produce the prototype with some or all functionalities.

- Step Two: Find out how and/or where to fabricate the prototype.

As listed below, several methods and places make it possible to make a prototype.

1. **Home-Made**

This choice will likely be the least expensive and most cost-effective if the essential equipment and technical expertise needed to create the product are available. Replacement raw materials like paper, plastic, cardboard, etc., can be used to create your prototype if your product is not very technical. Many excellent products have been built in a home office, kitchen, or garage.

2. **3D Printers or 3D Print Shops**

Three-dimensional (3D) print shops create a physical object from a digital design. The process works by laying down thin layers of material, in liquid or powdered plastic, metal, or cement and then fusing the layers together. The advantage is that the process produces a tangible prototype; however, it will have little functionality.

3. **Universities**

Engage with local universities and ask if they have a research department in your specific interest areas. After you find out the places that the university majors in, such as mechanical engineering, food science, textile and design, information systems, etc., you might approach them for support. Since 2018, most universities have created fabrication spaces within their research buildings where you can get help or use them to produce your prototypes. In addition, some universities have allotted budgets for community engagement, which means that universities may be able to construct specific prototypes for little or no cost. Even if the university cannot fully make the prototype, they might offer support or collaborate on the project.

4. FabLabs

More and more fabrication spaces are appearing every day worldwide, especially in big cities. A FabLab or fabrication laboratory is a small workshop offering various personal fabrication services. They are typically equipped with an array of flexible, computer-controlled tools that cover several different length scales and various materials to make new designs. They are an excellent way to get a prototype fabricated.

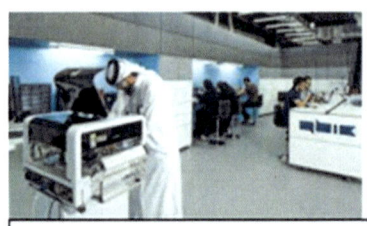

Qbic Fablab is located in the UAE.

Generally, FabLabs are free to use; however, there are charges for time and/or use of equipment. On the other hand, regional Fablabs have membership fees before you can use them. They commonly have the leading equipment you will need to produce your prototypes (3D printers, CNC machines, laser cutting, electronic kits, and others), although most do not have state-of-the-art technology for your prototypes. They also offer support, training courses, and other services.

Flexible manufacturing equipment within a fab lab can include:

- 3D printers
- Microprocessor and digital electronics design
- Laser cutter, plasma cutter, water jet cutter, knife cutter
- Electronic workshop and software development for fabrications
- Milling or turning machines for fabrications
- Circuit board milling or etching

5. Freelance and Professional Prototype-Makers

Freelance and professional prototype makers are excellent choices for creating a prototype, depending on the level of complexity of the invention or whether a design must be made from scratch. If the vision is complex, an inventor can easily research a 'design fabricator' in a search engine and obtain a list of services (www.guru.com or www.upwork.com).

6. Professional Services in Specific Areas

If your invention is very complex and there is no one expert in what you need to build, you may want to hire experts in several areas. Using search engines, inventors may find a list of companies specializing in several technical areas, like www.thomasnet.com and www.Alibaba.com, which post lists of suppliers. These sites are good options if you make a more mature prototype closer to mass production. Inventors may find suitable suppliers for several different types of technical prototypes. Some of them offer design services as well as fabrication services. The total cost of the prototype will be much higher than the options mentioned earlier, but the level of functionality in prototyping will be unmatched.

There are several options for prototype design and fabrication. Depending on the complexity of the inventor's idea, one could avail of a plethora of assistance matching your skill level and budget. If the concept is new and valuable, make sure to have the company sign a non-disclosure agreement so that your supplier does not become your competitor.

Case Study

Case – From Cancer Research to Hair Treatment

By Chara Lee-Brinson
Founder O'Hara Organics

O'Hara went from a late-night brainstorming session to a

prototype/product within six months. That is not to say that sales or money or money was being made. All it means is that it had gone from an idea to a tangible product that one day had the potential to be sold and change women's lives.

Background: I could not afford to go to a salon, and a work colleague told me I needed to start taking care of my hair. My husband was diagnosed with a rare cancer earlier, so we began researching ways to keep him healthy during his battle with cancer. This led us to practice a more holistic lifestyle focused on herbs and natural healing. To help my husband stay healthy, we researched herbs from around the world and their healing properties. My hair was dry and unmanageable. There were countless nights spent researching to solve my problem. I read article after article on what people should do to their hair to tame it, make it shine, or even get it to grow.

Research: What my study showed was that I had low-porosity hair. Something that is rarely identified and talked about in mainstream hair products. This meant that my hair had tightly aligned cuticles, keeping moisture from entering. Some articles even say that low-porosity hair can be water-resistant, meaning it is difficult to get water and products inside the hair instead of just sitting on top. Therefore, the proposed solution was that I would need oils with nanoparticles to penetrate my hair and produce the results I sought. Armed with that knowledge, I set out to find the oils that would work best for me. Within weeks, I had a group of 13 oils that did what I wanted. It made my hair manageable; I did not get the frizz after workouts, and my hair would hold a style for a week at a time.

Design: With this new miracle oil, the next question was, did I want to share this product with others? And if so, who was my target audience? How would I package it? What would defining characteristics make my product stand out in the marketplace?

Prototype and Testing: We concluded by making C'Hara for people like me, so I was going to start small with just friends and do a word-of-mouth campaign, especially since making this product was not my regular job but more of a hobby. In the end, O'Hara is an organic boutique product with a glass bottle and dropper. 'A little goes a long way' was coined. Unlike boutique products, the price point is attractive to my consumers because the small bottle lasts up to two months.

Case Questions

1. What was the initial motivation for the idea?
2. After the idea, how did the creator collect the information?
3. What were some of the personal issues that prompted this case author to initiate her business?
4. In this case, did Chara imply who the target consumer is? If so, briefly describe the target consumer.

Chapter 4

4. Packaging and Logo Design

Another essential part of production is creating the aesthetics of the product. It is no use having the most incredible effect worldwide if the packaging does not reflect the contents. This section is exciting because the packaging and branding of the logo are your way of summarizing and highlighting your product or service's value to your specific customer. Before a consumer buys a product, they probably study the inventor's online representation or packaging of his product or service.

Packaging is the part that surrounds the product and is the creator's direct visual message to the public about the purpose, value, composition, container, or wrapper of a product. In the packaging design, the inventor may consider function, aesthetics, language, and other relevant data influenced by market needs.

Package decisions lead to branding. Branding is defined as the way of identifying your business. It is how your customers recognize and experience your business. Companies with a good brand image will attract more attention from potential consumers to perceive the products' quality. Brand perception is the level of emotional bond between your business and the consumer. This is where logos are essential in creating a brand image for your product. A consumer will associate the quality of the product with the logo of the company developing an opportunity to market with minimal explanation.

A. Packaging Overview

Packaging is the cover of your product/service, whether physical or online. For brevity, this section will only discuss and make recommendations for physical products and will not elaborate much on online branding. With that stated, as your idea gets to the point of opening to the public, the creator must keep a few things in mind:

- Your product packaging and your online branding should have synergy. Make sure you match packaging, web, and social media later in the start-up process.
- When designing packages, always consider the function of packaging on your website and social media: color palette, fonts, or any visual aesthetics connected to your business.
- Make sure that packaging reflects your company's mission, value strategy, and where and to whom you desire to sell your goods.

I Generic Steps for Packaging

Providing specific packaging strategies for products and services across all industries would be impossible. Below is a simple recommended process that entrepreneurs may use as a checklist for completing packaging. The tips are not specific to a field or industry does not include tech or any types of services.

1. **Establish Your Value and Message** – Once you define the product or service's value, strategy, and how they compare to competitors' products or services, consider where your product will be sold and how it appears on the shelves or next to the competitors in general. Suppose the competitor's packaging is not very clear and straightforward, and all the products in your niche are packaged too simply. In that case, you may develop a colorful visual or luxury materials, so your design stands out. Your package should answer the questions.
   a) What is the product?
   b) Who is the consumer?
   c) and How are you selling the product?

2. **Develop a Packaging Budget** – Package costs are separated into two parts.
   a. Onetime costs will be the cost of designing the package, including logo design work, template purchase for the work, etc., to create a package design ready for the printer.
   b. Per-item costs – is the cost of labor and material used in the package.

You will need to estimate how much you are willing to spend since the packing material will affect the product price structure. Prices can range from extremely cheap to super expensive, but consider that cheap packaging may not be the better choice in the end since the presentation and appearance are as important as cost control. You must decide the type of packing material most suitable for marketing and transporting your product (i.e., corrugated board, plastic, papers, padded wrap, etc.).

3. Generic Versus Custom Design Decision

There are three basic choices for a creator's packaging decision, and they all have very different budget impacts. These choices are custom packaging, customer graphics, and generic packaging.

> ***a. Custom packaging*** – A collaboration process starts with the creator and a graphics designer, where they summarize all the details, product description goals, market communication goals, and overall message for the product or company logo. The  logo should answer the question, 'Who is the consumer?' A boomer may respond to luxurious materials and prominent logo placement. A more rustic and discreet logo may attract Gen Z. However, remember that the beauty of the packaging will not matter if the cost breaks your

packaging budget. This option is usually expensive.

b. ***Custom logo and décor on generic packaging*** – Generic packaging can promote the brand's identity by being customized
with a logo and/or other design elements that can create a distinct impression. Small package changes and a powerful logo can make a generic packaging kit, essentially only functional, appear more unified, and advertise your unique brand. A branded package needs to be simple and have a straightforward logo design. The overall cost is much less than custom packaging.

c. ***Generic packaging*** – is readily available packaging that is easy to use right away with no customization, and is usually used for bulk shipments. It is the fastest
way to get your product into the customers' hands and is the most economical decision.

II Logo and Packaging Draft

One of the most creative processes in inventing a product is logo and packaging drafting. The whole procedure of creating a package is collaborating with experienced graphic designers, marketers, and the creator. The team summarizes the many points (target market, product values, company strategy, competitive information, necessary logistics, etc.) to create the desired message to the consumers and public through logos and packaging.

Ideas for visuals, fonts, colors, textures, language, sizes, and drawings are developed for your review in this creative phase. The team considers the following, but is not limited to:

- How will the package be filled?
- How will the container be sealed and unsealed?
- Quality of functionality and life cycle of the package.

After several design packaging and logo renders are developed based on the above considerations, the team enters a final step. Although the packaging choices may vary in look and quality, the final decisions are often driven by the manufacturing cost of the package and the end consumer's value impression of the package style. Next, production quantity also influences cost (recall the term 'economies of scale'). Customized corrugate and plastic packaging will always have higher minimum production requirements.

If you sell or ship products in a retail environment, you have to purchase UPC (universal package codes). Nowadays, sellers may also create a bar code with additional product information.

Some product models require part labeling. Depending on the country in which you plan to sell your products, you may have to label parts or ingredients. For example, in UAE or USA, all food products sold in a retail environment require nutritional labeling of the ingredients in the home country's language. For retail food items, depending on the product, nutritional facts and components must be obtained through that country's authorized food testing facilities. Next, when you provide text on the packaging, you must consider the pertinent information that you must provide to consumers and the government's disclosure information requirements. This will be an additional cost to your final prototype budget.

III. Developing a Mock-Up

Once the design decisions are finalized, you must create a sample mock-up, or at least a 3D mock-up, of your packaging artwork. Coupled with your product, your logo and packaging allow you to show potential investors, test consumers, and other stakeholders how the actual product package will look in real life. In developing aesthetics, remember that fonts and photos are integral in creating the design to produce a successful prototype package for illustration purposes.

The two types of mock-ups that you can produce are physical and digital.

a. Physical

Creating a physical mockup will require a professional photographer for the images needed by the graphic designer. A physical mock-up will enable you to

see the product's appearance and allow you to revise or modify the design aspects. A physical mock-up, along with a logo and a prototype, will always be more appealing to potential investors and stakeholders. It demonstrates that the creator has done the initial legwork and that their idea has actually come to life. Physical mock-ups are far more palatable for potential customers, in which opinions can be collected and analyzed.

b. **Digital**

As the name suggests, a digital mock-up is a computer-generated image rendering that allows stakeholders and test markets to see how the design would look. An advantage of digital mockups is that these visual renders may be sent anywhere electronically. This will allow creators to send the prototypes to many individuals and groups without the cost of shipping the actual models. After the feedback process, digital mock-ups will allow you to make changes quickly in any 2D or 3D graphic design software.

IV Package Marketing Test and Feedback

If your product or service is commerce-driven, you need to do some package testing market research for your creation packaging. This testing will help you create the right visual aesthetics, color, patterns, messages, and material to appeal to your potential consumers. This exercise aims to understand what users and potential consumers think of your prototype and package.

Case – Interview with a Graphics Expert

Case – Keys to a Professional Render

By Sharon Elizabeth Nakitto, Graphic Design Professionals of Dubai

Before rendering a piece of work, the preliminary steps are coming up with the design, shape, color, etc., required to have a video rendered. The first steps include:

RESEARCH

The particular topic, design, or project of your desire needs to be thoroughly studied to give you insights into its reason and purpose of creation. Ideas are generated through researching, and you are exposed to various types of information. All this is laid onto a piece of paper, and then sketching follows.

SKETCHING

Everyone who is a creator, investor, designer, animator, etc., is open to sketching. Customers who hire designers to build on their projects always show up with a rough sketch of what they would like to have, so sketching is not done only by professionals. When sketching, it is best to keep all your ideas on board because these additions and subtractions are not excluded in formulating opinions. The idea is to keep all that you have sketched for a future purpose.

MOULDING

This is where you transfer the sketch onto a computer by using tools in any desired software you're working with. While building or after completing your sketch design, molding helps you build up a design into a 3D module where you can see all corners, edges, front, and back of the design. With this, you can accelerate mold by addition and subtraction until satisfied.

TEXTURING

Texturing shows you how exactly you would want your ideas to appear physically. This is where folds, smooth and rough areas, spots, etc., are shown. With this detail, a picture is transformed into realism. When sketching, an idea is built up, but it does not really show how the final piece will look, even if particular areas are marked in different coloring to show the difference. Once you texture your work on a 3D module, it becomes more practical and efficient.

LIGHTING AND SHADING

The beauty of your work is brought out through lighting and shading. Most clients are interested in works that speak volumes about realism. With this, you can choose the correct correction of light or shade on specific areas and how it will appear to the viewers. From a physical point of view, a presentation of a piece of work is based on shading and lighting. So, when this is applied to your 3D module, it gives you a feeling of reality.

With all this completed, video rendering can bring your dream project to life even before it is made. With high-quality video and virtual reality, project video rendering helps you see exactly how your project will turn out to be once completed. A virtual walkthrough will show you how it will appear in reality, and with this, one can eliminate things that can't be seen in a sketch. Having all this completed will depend on the kind of project you are working on. Most works take a day to months just to have something satisfying. As a designer, setting a certain period for the projects I am to work on depends on how complicated the work is. With software, one needs to know what project they are working on to complete your designs. As a designer, I work with a wide range of software because each comes with unique features. These include;

CINEMA 4D

This software will complete all your desired projects without transferring your work to another software. It is well known for its versatility, as it works best on different kinds of projects like animations, architecture, logos, etc.

BLENDER

Best known as a beginner's choice. It handles many processes beyond design, such as rendering, motion tracking, and more. The robust software provides the best texturing and sculpting tools, including unique features.

ADOBE PHOTOSHOP

One of the most accessible and most commonly used software. It offers all 3D capabilities, including 3D printing, textures, and maps. This is a start-up kit for fresh designers. It also enables you to work with different software by transferring your project to another for more rendering.

These are among the many soft wares available. With these kinds of software, one needs to be familiar with them before attempting to use them. Taking a quick course or tutorials on how to use the software is good because no one can start using it right away.

As a designer, I took a two-year course and practiced even more to build on my skills. I have mastered the use of;

1. Cinema 4D
2. Illustrator
3. Photoshop
4. InDesign
5. Sketchup
6. Motion builder
7. Solid works

Case Questions

1. As a beginner in graphic design, what two software would you choose? Explain.
2. What are at least four areas that must be considered when developing a render for a prototype?

Chapter 5

5. Prototypes Testing

Prototype testing is the process of testing your prototype with real users to validate the design objectives of the creator before anyone starts investing, building, and selling. The feedback from the test will help the creator to identify problems, correct mistakes, and improve functionality. With that, the goal is to maximize the quality of the product or service and meet user needs and expectations. Investors like to see this process done because it implies that the consumer has seen that product or service and is willing to buy it. The risk is lowered for investors.

Once you have built your prototype and started testing, you must develop a list of clear goals (e.g., various functionality, perceptions, aesthetics, etc.) of the product/service you want to validate. This will help you define areas of the prototype you need to measure. At this point, it is time to try the prototype in front of a limited set of users.

There are three main ways in which to test a prototype.

1. **Focus groups** are a small set of six to ten people who usually share common characteristics such as age, background, geography,  etc. The set comes together to discuss aspects of your product or service with the team. These people should reflect the ultimate consumers to whom you intend to sell, i.e., your target audience. Let this group look at, touch, and test the prototype while answering a list of mostly open-ended questions. You must get results that reflect the consumer's reaction to various elements of your design and take in quick, detailed responses and excellent feedback to create a better product.

2. **Observation** testing is a technique that involves directly watching potential consumers in their natural or controlled environment. In this technique, information is gathered from the test group by the testers through an analysis of their behavior while using and observing the product or service in question. The creator's team will capture data directly from the users, and they will make notes and answer questions about the prototype. The observation technique is excellent for products/services in fashion retail, search and click technology, and artificial intelligence projects. For instance, watching how online fashion shoppers click on search sites, what they are drawn to on Google, and which keys/photos they go to after opening a web page.

3. **Surveys** allow you to get detailed feedback by using measurable scales. With this method, the creator develops a list of scaled questions regarding areas of the product/service in which he requires input, such as aesthetic appeal, perceived quality, likelihood to buy, comparisons to competitors, process appeal, willingness to pay a higher price, functionality, etc. If you have a tangible product or service, in-person interviews or face-to-face surveys can get you insightful and valuable responses from more test consumers. For intangible services, software, and information technology-related items, online surveys are accessible to any participant internationally or globally. Creators may create the same surveys online and distribute them via email or link.

F. Simple Steps for Testing a Prototype

Below are **four steps** to building and executing your prototype and packaging market test. Make sure to pick the right questions to ask your test group. Depending on your type of product or service, or process, you have to ask questions that will give you even more insight. Make sure you pick the right ones to get effective feedback.

1. **Develop Your Prototype Testing Tools**
    a. First, you need to draft a set of questions (i.e., open-ended, close questions, observation questions) relating to the areas you are testing or validating. This is where creators must articulate what goals he is testing.

Examples of overall goals:

'I want to find out if people can easily order and have delivered restaurant food through my website' by Talabat.

Or, 'We expect to see if users can navigate with ease through my app.'

      b. Next, pick the right questions to ask the test group. Depending on the industry, prototype questions should be customized and tailored to the creator's needs. Make sure to develop a list of the scaled focus group, observation questions, or survey questions regarding areas of the product/service in which he requires feedback, such as aesthetic appeal, perceived quality, likelihood to buy, comparisons to competitors, process appeal, willingness to pay a higher price, functionality, etc. Below are a few examples of test questions that apply to Talabat (an online food and grocery delivery service app).

- Have you ever purchased from other food delivery sites before?
- How easy was it for you to view restaurants?
- What did you think about the Talabat experience?
- How was the language used on this page?
- Can you tell us what you think of the restaurant's selection?
- What do you believe is the ideal price range for the service?
- What is your opinion on the quality of the overall website?

- How likely would you recommend our service to a friend or colleague?

Ensure you avoid being vague with your end questions to get those key results relevant to your goals.

2. Choose the Right Audience

Whether selecting participants for a focus group or observation, the person you invite to test your prototype should have a minimum knowledge or interest in the product. If you were launching a fitness-tracking app, it would make sense to only include test participants who work out at least twice a week.

Two types of questions will help you find out more about your participants.

a. Demographic usability testing questions

Collecting demographic information will give your results more context, allowing you to spot usability trends across different demographics, such as ages, nationalities, genders, income groups, etc. You have to word them carefully to avoid sensitive topics. Here are a few examples of well-worded demographic questions to ask as part of your user research. Asking demographic questions avoids making assumptions about people by mistake. It also gets you the context you need – without making people uncomfortable with overly direct or specific requests for personal info. A positive start will help your participants feel relaxed when taking the test.

- What age group are you in? 18–24, 25–30, 31–40, etc.
- How do you describe your gender?
- What is your relationship status?
- What is your household income? (Provide ranges as options).
- How do you describe your ethnicity?

b. Background questions

Besides demographic information, asking questions about the user's habits and preferences is also helpful. For example, is the test subject using your competitor's product and is he comfortable with the product?

- How often do you use our product?
- Which features do you use most?
- What type of product do you usually use?
- Have you used any of these products before?

Asking the right demographic and background questions before the test will help you choose the right test participants.

3. Execute Your Experiment

It is prototype testing day, finally. Do a trial run with a colleague or friend if the testing is in-house, or set up a pilot test from your remote testing app so that you're 110% prepared before a real-world test.

4. Collect and Analyze Consumer Data
a. Organize the issues

The first step to understanding the data you have collected is to organize the issues you recorded during usability testing.

The simplest method is to write down all the issues along with a complete description of where in the design it happened, how it occurred, and the task the user was performing, plus all the details describing the problem.

Alternatively, arrange the data in a table format according to one of the methods presented by Lewis and Sauro in Quantifying the User Experience: Practical Statistics for User Research. On one side, include each task with its separate issue. On the other hand, include a list of all your usability test participants so you can mark down every user who has experienced the same problem.

b. Prioritize issues based on criticality and impact

Categorize the problems based on the severity of the issues identified by ranking your findings. This will help you to classify the problems you need to handle according to their priority. This exercise will make it easier to allocate your resources and effort to tackle urgent issues according to their severity.

For instance, a usability issue that makes it hard for users to find the call-to-action on the page might be a more critical issue than a typo.

We can categorize usability issues based on five severity levels (NN Group):

4 – Critical: Write down all problems that interfere with the user experience and impede users from completing tasks. These issues need to be fixed on a priority basis before a release.

Examples: The user did not receive a confirmation message after making a payment, or they cannot sign up for an account.

3 – Serious: Include issues that have to be fixed as soon as possible as they slow down a user's experience with your website or product.

Examples: The user cannot reset their password.

2 – Medium: Includes issues that might not interfere with the user experience often but can still be frustrating for users to experience.

Examples: The user has to scroll a lot to find a category in a dropdown, or the text is too small on the pricing page.

1 – Low: These are superficial issues necessary to fix but do not affect the user experience. Issues like spelling errors or unclear images might affect your brand image, but do not severely affect your UX.

Examples: a misspelled word in the heading or a logo that is not updated on the About Us page.

0 – No issue: This situation occurs when a user reports a problem that turns out to be a non-issue or a feature request, so you don't need to take further actions right now.

Chapter 6

6. Intellectual Property Phase Overview

The definition of intellectual property is a category of property that includes the intangible creations of the human mind. This property, i.e., ideas, theories, inventions, innovation, writing, music, and art, is protected by copyright and patent laws. Intellectual property laws in most countries usually cover the creation of a new product or idea and define the activities under the control of the patent or copyright holder and those that infringe on his right. This gives people control over the information or product they have created and prevents copying, which incentivizes more innovation.

Since most of today's intellectual property is computer-based, a few extra dilemmas arise due to the escalation of the use of artificial intelligence. The main concern is how to handle creations invented by AI without the involvement of a human being. For example, AI's ability to come up with music that may or may not be based on an existing theme. For example, Sony Computer Science Laboratories' Flow Machines platform offers a song-co-writing application that can compose new music based on thousands of existing music samples. This application is now widely used by composers, which may lead to ownership conflict and a loss of copyright revenue to the original composer of the music sample.

A little bit of history: The UK laid the basis of patent law and copyright with 'The Statute of Monopolies (1624)' and the 'British Statute of Anne (1710),' establishing the concept of intellectual property. Therefore, it is only fitting that the UK was the first country to provide copyright protection for 'computer-generated' works with the Copyright Designs and Patents Act 1988. The law states that in circumstances where an otherwise copyrightable piece is created but no natural person qualifies as an author, the 'producer' (human) of the work is deemed the author.

In the UNITED ARAB EMIRATES, copyrights are protected under UAE Federal Law. The UAE is a member of several WIPO-administered treaties, including the Berne Convention, the Rome Convention, the WIPO Copyright Treaty, and the WIPO Performances and Phonograms Treaty.

A. How to File a Patent in the United Arab Emirates

Patent protection was available until January 2021 by filing a national UAE patent application with the UAE patent office, through the Patent Cooperation Treaty system, or by filing a patent application with the regional Gulf Cooperation Council (GCC) Patent Office in Riyadh. However, the system is undergoing legislative reform, and new patent applications are no longer accepted.

You can register a patent in the UAE online on the International Centre for Patent Registration (ICPR) website, which is administered by the Ministry of Economy and follows the process below.

1) Fill in the application form through the system. The first step of patent registration is filling out and submitting the patent application along with the complete specifications and details of the invention, listing the elements that need protection. Also presented are illustrations, a summary, bibliographic information, an abstract, Patent Cooperation Treaty (PCT) documents, an extract from the partnership or commercial registry, a power of attorney, Emirates ID, an undertaking related to the application about making all the records available, and more.
2) Payment of service fees
3) Verification and legal examination of the application and submitted documents by the department's employees. A legal and formal investigation follows the application's acceptance after submitting all the documents. The average time for this process is two years.
4) Notifying the applicant of the outcome of the formal and legal examination of the application through the system.
5) Invitation to the substantive examination. The patent then undergoes scrutiny per the obligatory criteria and norms, and more documents and information may be requested.
6) Request substantive examination and payment of fees.
7) Examination by the technical examiner

8) Examination report issued by the examiner
9) Payment of publication fees
10) Patent registration
11) The publication will be released if the MOE accepts the application. The patent is published in the official UAE Gazette along with additional information and is open to comments and objections.
12) Certificate issuance will depend on the successful defense of any comment or objection raised to the patent.

The documents to be submitted to the Ministry of Economy for natural persons residing in the UAE are as follows:

B. General Mandatory Documents Required

1) Bibliographic Information
2) Description (Arabic or English)
3) Application Form
4) Description (Arabic or English)
5) Claims (Arabic or English)
6) Description (Arabic or English)
7) Abstract (Arabic or English)
8) Drawing (Arabic or English)
9) Representative Image – Minimum 500x500 pixel
10) Certified copy of priority application
11) PCT Documents
12) Extract from commercial register or partnership

13) Power of Attorney
14) Emirates ID
15) Patent or Utility Certificate Documents Receipt
16) Assignment
17) Extract from commercial register or partnership
18) Commitment to submit documents belonging to application
19) Industrial Design or Drawing Documents Receipt
20) Sequence listing
21) Proof of applicant types

The patent registration process is complex, time-intensive, and expensive, but it is worth protecting your invention. If two years is too long to wait, you can register your copyright as a work of science within a month. This method protects new products, inventions, and scientific works and involves searching national and international automated databases for existing registered work.

It will grant its owner exclusive rights like:
- Right for commercial usage
- Reproduction rights
- Right of creating derivative works
- Right to public display and performance
- Distribution rights

Chapter 7

Although thousands of ideas creators have daily, each may be illustrated or visualized in a way that suits your needs. One may possibly need a tangible image for several reasons. From detailed drawings for an application designer on a project to food samples for taste testing in a mall to a high-fidelity prototype to present for investor consideration, there exists a unique and appropriate way to convey all parts of your initial idea. Furthermore, each type of idea requires a specific prototype. Below is a helpful chart showing various kinds of innovations. Each innovation illustrates its definition, a historical example, and a recommended prototype approach.

Invention Area	Recommended Prototype(s)	Definition and/or Examples
Products		
Food/Beverage Innovation	Initially, you may create drawings and storyboards and include potential logos and products. Second, create trial product samples and/or working prototypes. Create several types of samples, so your consumer may try the variety.	**Definition**: Food innovation is developing and commoditizing new food products, processes, and services. Inventors create ways to make healthy, nutritious products that are accessible, unique, and/or sustainable. Example 1: The creation of plant-based, animal-free products Example 2: Plastic-free and smart packaging Countries in the EU were some of the first to ban single-use plastics. Many countries around the world have turned to edible and biodegradable packaging. The edible packaging uses food-grade polymers, such

		as seaweed, which are safe for consumption.
Restaurant Innovations	Initially, you may create drawings and/or storyboards of your concept and layout. Use wireframe prototypes to illustrate location layout and a building. Feasibility Prototypes for software retailing to restaurant software projects	Definition: Bringing customers value to restaurants through channels of services such as identity, appeal, aesthetics, technology, community, customer experience, altruism, commonality, simplicity, avoiding the hassle, saving time. Three examples are online and mobile ordering, digital point-of-sale systems, waitlist, and reservation management.
Technology		
Technological Innovation (Hardware)	Initially, you may create drawings or storyboards of the concept, value, and layout. Next, low-fidelity and/or video prototypes are suitable for a general example. Last, it will be best to develop a high-fidelity prototype before	The definition of technological innovation is new inventions derived from research developments, technical knowledge, and tools independent of product and service. An example is a plug-in device or a built-in microphone with voice

	releasing your idea for feedback or investors.	recognition available with the computer device.
Application / Software Innovation	Initially, you may create drawings and a storyboard of the software or app's concept, value, and layout. Later, you might develop a low-fidelity application prototype for limited trial and feedback. Video prototypes explain clearly to the app developer in stage two. Last, you might strongly consider developing high-fidelity prototypes before releasing them for critique or to potential investors.	Definitions: The generation of new ideas for programs and other operating information used by a computer. Software innovation is the creation beyond the schedule and additional operating information used by a computer or phone. They are discoveries beyond simply working to generate new ideas. Example 1: Computer vision is, in computer terms, 'vision' that involves systems that can identify items, places, objects, or people from visual images and those collected by a camera. Its software allows the smartphone camera to recognize which part of the image it's capturing is a face. Example 2 is autonomous driving software by Tesla.

Improvement Innovation	Start with drawings, cartoons and storyboarding, low fidelity prototype, high fidelity prototype, video prototypes to illustrate your ideas and how they are different/add value to existing products and services.	Definition: Create something by making it more effective, healthier, faster, more efficient, easier to use, serve more purposes, longer lasting, cheaper, more ecologically friendly, or aesthetically different, lighter weight, more ergonomic, structurally different, with new light or color properties, etc. Example: Guitar with a battery-operated auto-tuner
Mechanical Engineering	Start with creating drawings, paper rendering, or storyboarding of the concept, value, and layout. Use a Teardown of existing products and draw creative changes that you desire. Video prototypes are recommended to explain to the fabrication group. Low fidelity prototype or high fidelity prototype for later	Definition: Mechanical engineering innovation is a creative function or style and change that can radically evolve from original principles in physics, process, and mathematics with materials science to design, analyze, manufacture, and maintain new mechanical systems.

	product development and testing	Example: The first sea-bound floating rollercoaster
Services		
Supply Chain/ Logistics Innovation	Initially, you may create drawings, Paper Rendering or storyboarding of the concept, value, and layout 1. Low-fidelity and/ or video prototypes prototype for general example high-fidelity prototypes before releasing to market CAD or 3D models can be effectively used to convey your idea.	Definition: Logistics innovation refers to new technology, services, processes, and ideas used to improve logistics operations (Scott, 2009). An example is the Internet of Things (IoT) connection of physical devices that monitor and transfer data via the Internet without human intervention. IoT in logistics enhances visibility in every step of the supply chain and improves inventory management efficiency. Integrating IoT technology into the logistics and supply chain industries enhances efficiency, transparency, real-time visibility of goods, condition monitoring, and fleet management.
2. Re-Envisioning	You may start by creating drawings,	Definition: Inventions that become simpler

	cartoons, and/or storyboards to illustrate what your product or services do. Explain the transformation and value that you cause. You may create levels of semi-functional samples and/or working prototypes.	and more practical may expand or even transform into something totally different. Examples: *Teflon material – used for cooking pan.*
Consolidation and Delivery Innovation	Nowadays, there are several layers to a consolidation/delivery idea. Like Talabat or Door Dash, there are layers of creation: Application prototype (low-fidelity software prototype), logistics drawings/map, essential equipment, and service center (drawing or rendering); an example.	Definition: Invention may become more straightforward, more practical, it may expand, or may even transform into something totally different due to using tech applications to change a face-to-face business into delivery initiatives. Example: Amazon and noon have standards: The business is built by (1) developing software from suppliers who make popular products (2) Creating a payment portal, and (3) developing a logistics and delivery system to customers for geography.
Social Media	You may start by creating drawings or a	Definition: system for disseminating

	storyboard to illustrate what your application does. Explain the value that you cause. Later, you should develop a low-fidelity application prototype for limited trial and feedback.	information over the Internet to a selected group of followers. Social media platforms are used by people to publish their daily activities, comments, and photos, as well as re-publish information posted by others. Example: The creation of Tik-Tok.
Healthcare Innovations	These could be products or service innovations. Both start with creating drawings or storyboards to illustrate what your application does. Explain the value that you cause. For Services, higher levels of rendering must be developed with your idea for added value.	Healthcare innovation may be defined as any simple or complex developments that lead to improvements in health outcomes and patient experiences. Example: The Sterile Processing industry for surgical equipment.
Fashion Innovation	For fashion, you may create drawings and/or storyboards while explaining the value of the aesthetics, textile, and product details. Create product samples and working prototypes. If there are logos, illustrate them as well. Create several	Definition: Fashion innovations are form or function, or style and can be radically evolved. It involves the breaking down old ideas or incremental evolution involving the evolution of new opinions from old ideas.

	types of models so your consumer may try the variety.	Example: X suits – new stretchable material
Arts Innovation	Similarly, in art, you may create drawings and storyboards while explaining the value of the aesthetics and unique product details. If you are doing innovation, use a product sample demonstrating your pieces' uniqueness or added value to the industry (e.g., Digital art).	Definition: Artistic Innovation is generally defined as the introduction in the field of something new (Cloake, 1997) Example: The creation of moving digital paintings
Humanitarian Innovation	This is a critical area of Innovation but might be complicated.	Definition: Creating programs and organizations for the betterment of society. Documents like the Universal Declaration of Human Rights. Organizational innovations like the Boy Scouts, the Red Cross, and the Olympic Games exist. Example: Creation of the World Health Organization (WHO)
Gaming Innovations	Initially, you may create drawings and a storyboard of your game concept's	Definition: Gaming innovations are new discoveries in running specialized applications known as video games

	concept, value, and layout. Later, you might develop a low-fidelity application prototype for limited trial and feedback. Video prototypes explain clearly to the app developer in stage two. Last, you might strongly consider developing high-fidelity prototypes before releasing them for critique or to potential investors.	on game consoles, personal computers, and online gaming. Example: The Grand Theft Auto III Platform for PlayStation 2
Agricultural Innovations	Initially, you may create drawings and/or storyboards of your product idea. You may also have a service or environmental idea. This is where you may express the details and the values of your innovation in video rendering; CAD or 3D models can be effectively used to convey your innovation concept. If it is a product, create a product sample and working prototype.	Definition: Agricultural innovations are primarily concerned with satisfying the need for increasing production (of food, fodder, secondary products) and enhancing quality (of produce, production process, and growing conditions). Example: Automated feeder systems that provide animals with feeding mixtures tailored to their specific needs and in the right amount.

Since there are thousands of types of innovation areas, the above chart is nowhere near complete or exhaustive. However, this is clearly directional and may provide insight into how to illustrate the ideas that were previously in your mind. Whether you wish to monetize your vision, improve an existing product/service or develop a concept that will enhance the environment, you might now have a tool that will better illustrate what was previously a brainstorm.

Please remember complexity needed in your prototype heavily depends on your purpose in terms of who needs to see it. Consider, are you prototyping for the purpose of:

- Trial to your friends and family
- Practice for a critique by a small personal group
- Trials for choosing a few of many of your initial ideas to a limited test sample
- The initial investment in developing a high-fidelity model
- A significant investment by a bank or private group based on your final model or prototype

Along with a great business plan, a proper prototype exponentially increases your opportunity for success. If your initial model is not well received, don't give up. Collect and analyze feedback from your various groups. Plan the changes, develop a better prototype, drawing, or model based on the feedback that you received, and retry your newer version. Remember that a great quality product and service development project involves a few key steps: plan, do, check, act, analyze, and plan again. To conclude, a famous 1990s

American film expresses an appropriate and inspirational saying, 'If you build it, they will come.'

Bibliography

(n.d.)

Benedetto, D. Z. (2010). Radical Fashion and Radical Fashion Innovation. Journal of Global Fashion Marketing, 1:4, 195–205.

Butcher, K. (2006). earning from text with diagrams: Promoting mental model development and inference generation. Journal of Educational Psychology, 182–197.

cal-international.com. (May 2022). 10 Mechanical Engineering Innovations That Helped Define Mechanics As We Know It Today. Retrieved from cal-international.com: https://cal-international.com/2021/05/06/10-mechanical-engineering-innovations-that-helped-define-mechanics-as-we-know-it-today/#:~:text=Mechanical%20engineering%20is%20a%20branch,manufacture%2C%20and%20maintain%20mechanical%20systems.

Doty, R. &. (1997). Commonality of processes underlying visual and verbal recognition memory. Cognitive Brain Research.

Finlayson, K. (July 29, 2020). Prototype testing: How to nail your next product launch. Retrieved from https://maze.co/blog/prototype-testing/.

foodsafedrains.com. (March 2022). 8 INNOVATIVE FOOD TECHNOLOGY TRENDS FOR 2021. Retrieved from foodsafedrains.com: https://foodsafedrains.com/blog/innovative-food-technology-trends

Indeed Editorial Team. (July 7, 2022). 10 Types of Prototypes (With Explanations and Tips). Retrieved from https://www.indeed.com/career-advice/career-development/types-of-prototyping

Inkbot Design. (October 9, 2020). 5 Packaging Design Tips for Better Branding. Retrieved from https://inkbotdesign.com/packaging-design-tips/

Innovation, B. o. (2022). What healthcare needs next The future of innovation starts with a vision for better healthcare systems. Retrieved from https://www.boardofinnovation.com

insights, S. (March 2022). Top 10 Logistics Industry Trends & Innovations in 2022. Retrieved from Startus insights: https://www.startus-insights.com/innovators-guide/top-10-logistics-industry-trends-innovations-in-2021/

Keller, D. (June 2022). Follow This 10 Step Process To Create Packaging Design That Sells Your Product. Retrieved from https://blog.catalpha.com/packaging-that-sells-starts-with-our-professional-packaging-design-process.

Ltd, D. R. (July 2020). Observation Market Research. Retrieved from http://www.djsresearch.co.uk/services/service/Observation.

Marr, B. (September 30, 2019). The 7 Biggest Technology Trends In 2020 Everyone Must Get Ready For Now. Retrieved from https://www.forbes.com/sites/bernardmarr/2019/09/30/the-7-biggest-technology-trends-in-2020-everyone-must-get-ready-for-now/?sh=432f44162261: https://www.forbes.com/sites/bernardmarr/2019/09/30/the-7-biggest-technology-trends-in-2020-everyone-must-get-ready-for-now/?sh=432f44162261

Prototype Testing: A Step by Step Guide (2022). (July 1, 2022). Retrieved from https://qualaroo.com/blog/step-by-step-testing-your-prototype/

Qualtrics. (2022). Most common types of surveys available for market research. Retrieved from https://www.qualtrics.com/uk/experience-management/research/types-of-market-research-surveys/?rid=ip&prevsite=en&newsite=uk&geo=AE&geomatch=uk.

Richards, R. (January 26, 2022). Agriculture Innovation: 10 Tech Trends to Watch in 2022. Retrieved from https://masschallenge.org/article/agriculture-innovation

Rosencrance, L. (2021). Definition software. Retrieved from https://www.techtarget.com/: https://www.techtarget.com/searchapparchitecture/definition/software

Shuhalii, A. (August 2020). What is the difference between low and high-fidelity prototypes? Types of prototypes. Who is responsible for creating them? Retrieved from https://bootcamp.uxdesign.cc/what-is-the-difference-between-low-and-high-fidelity-prototypes-b1f3612f85f7

Thrive Wearables. (January 29, 2020). Rapid learning and risk reduction with Feasibility Prototyping. Retrieved from https://www.thrivewearables.com/rapid-learning-and-risk-reduction-with-feasibility prototyping/#:~:text=A%20Feasibility%20Prototype%20allows%20the,featured%20representation%20of%20the%20product.

Upadhyaya, V. (September 15, 2015). How to raise funding for a prototype. Retrieved from https://www.entrepreneur.com/article/250224

Veen, M. v. (2010). Agricultural innovation: invention and adoption or change and adaptation?, World Archaeology, 42:1, 1–12.

Will Keyser – In Benefit Venture Blog, I. (2022). Insight. Retrieved from https://venturefounders.com/startup-storyboarding/.